of the
German Schnellboot Forces

1935-1945

By: Richard W. Mundt

Cover photo: S-11 at high speed. Depending on boat type and engine work, the German E-Boats (Schnellboote) could reach speeds up to 43.8 knots.

This book is dedicated to
the Leader of Schnellboot Forces,
Kaptitan zur See and Kommodore
Rudolf Peterson
(1905 - 1983)

First published in the United States of America in 2010 by Crimson Books, Inc.
440 Thomas Avenue, Forest Park, IL, 60130, USA
www.crimsonbooksinc.com

ISBN: 978-0-9841652-1-6

Edited by Lynda Fitzgerald
Photos, unless otherwise noted, are from the author's personal collection.
Printed in the U.S.A.
Copyright ©2010

A very worn and tattered Reich war flag on this type 30 boat. The signalman is positioned on his special platform at the highest point on the boat.

Table of Contents

Acknowledgements

During the writing of this book, I found that more questions were raised then I managed to answer. Perhaps, with time, others will fill in the blanks. I extend a special mention of thanks to David Krakow for his contribution of unpublished photographs. Other individuals who contributed to this book, listed in alphabetical order, are:

Roger J. Bender

Hugo Berger

David L. Krakow

Richard Mundschenk

Fkpt a.D Karl M. Scheuch

Heinz Hildebrand

David Blake

Jerry Dutscheck

Richard Herman

Karl Kilb

Notice the bow of this type 38 boat low in the water. Towing the S-113 out of the operational area to Rotterdam, November 1942.

This unidentified type 38 boat is painted in a seldom used camoflage scheme.

Introduction

The development of the first German torpedo in 1879 and the first Imperial German Navy steam torpedo boat in 1880 would lead towards finding the most efficient use of this new weapon. The concept of using small, very maneuverable boats was tested during World War I by the Germans. At the outbreak of World War II, the German Navy had developed large, yet very maneuverable motor boats carrying up to four torpedoes.

As a young man in the 1970's my interest began when I was impressed by the sleek lines of these boats with their white painted hulls which gave them elegance as they rose slightly out of the water when traveling at high speeds. At that time very little information was available about these agile boats. The standard model kit at the time was of an early type of E-boat in 1:72 scale. My interest was rekindled in the late 1990's as the exploits of these boats started being published. The surprise gunfights with British MTBs in the channel, the capture of Venice without a shot being fired, the mass evacuations of troops and refugees from the advancing communist armies in the east. Enticed by information I obtained from veterans, period photo albums and individual photographs, I was compelled to turn this information into a book to share with other collectors of naval militaria and scale model builders. This work includes numerous candid shots of the different crews, most never before published.

When using this book one should consider all the standard Kriegsmarine uniforms, European and tropical. Idiosyncrasies of a specific flotilla, if and are mentioned in the unit history. The illustrated career and specialist training insignia are intended to illustrate only insignia worn by the men who served on the boats. Higher echelons are not represented.

Without a written record, it was difficult to attribute some insignia to a particular boat number, and I did not want to speculate. Boats being transferred, new commanders, or simply repainting at dockside often resulted in variations. There were also periods where a boat would have no insignia for whatever reason. The line drawings are to no particular scale and serve only as a backdrop for the insignia.

SCHNELLBOOT CAP BANDS

In 1932, the 1st Schnellboot half flotilla was formed and assigned the station tender Nordsee. The Nordsee served as the half flotilla's tender until the addition of more boats in mid 1935 when the 1st Schnellbootshalbflottille was retitled 1. Schnellbootsflottille and assigned to the newly built fleet tender Tsingtau. This was a camouflaged name to hide the ship's true mission and not come in conflict with the heavy constraints on Germany due to the Versailles Treaty. The Tsingtau's crewmen were to wear the cap band of the 1. Schnellbootsflottille.

In August 1938, the 2. Schnellbootsflottille was formed. This was the last Schnellbootsflottille to have a named cap band.

For a very brief time, there was a cap band Schnellbootsbegleitschiff Tanga, which translates to Schnellboot escort ship Tanga.

In November 1939, impending war led to the restricted wearing of named cap bands in secure areas, or shipboard for security reasons. By mid 1940, only the Kriegsmarine cap ribbon was authorized for the duration of the war.

There were two types of manufactured cap bands during the war. The early cap bands were of black artificial silk with the letters being comprised of gold plated copper wire. These tend to darken with age and exposure to humidity, sea water etc. The later war cap bands were entirely made of artificial silk with machine woven lettering.

The various types of Kreigsmarine and Schnellboot related cap ribbons. Note: a variation of the 1. Schnellbootsflottille exists with an additional period after the word Schnellbootsflottille.

The three patterns of metal cap eagles and cockades. Top center: Pre-war, early cap eagle. It is styled more like a party eagle. Right side: combination one piece insignia. Left the standard two piece type.

Previous Page

Top Left: A member of the 1. Schnellbootsflottille. Of interest is the variation cap band with the additional period after the word 'Schnellbootsflottille'.

Top Right: This sailor sports the cap band of the 2. Schnellbootsflottille.

Lower Left: S-boot cap band in wear with the summer uniform. Note the early pattern eagle and gun chief for small vessels specialist patch on his lower left sleeve.

Lower Right: A very dapper machinist/electric motor specialist of the 1. Schnellbootsflottille.

After the war, the early Bundesmarine utilized cap bands with Gothic lettering. This tradition was abandoned to be replaced by capital case Latin lettering. The term 'Schnellbootgeschwader' is a post World War II designation. The East German Navy (Volksmarine) made use of Gothic lettering on its sailor's capbands.

This Bundesmarine Sailor wears his Schnellbootgeschwader capband with his white summer uniform and fall winter dark blue uniform. (Photos courtesy of Karl Scheuch)

SCHNELLBOOT CREW COMPOSITION AND DUTIES

The Schnellboot crew composition varied by; type of boat, number of guns and mission. One example of a crew listing from a mid-size boat, S-30 type, is derived from the casualty list of S-35 and is as follows.

1 Commanding officer/usually Lieutenant zur or above. (Note: Occasionally senior NCO's were boat commanders;)

1 Senior NCO/acting as second in command; x1 Senior NCO/ acting as lead machinist;

1 Obermachinist mate;

1 Machinist mate;

6 Enlisted machinists;

2 Coxswain/Helmsman (Steuermann);

3 Enlisted radio operators/who doubled as hand signalmen and lookouts;

1 Enlisted torpedo mechanic/ who serviced the torpedo and smoke laying systems, served as boat's cook;

5 Enlisted seaman/boat operations, lookouts and gun crews;

1 Mine technician (not listed on casualty list of S-35).

As a practice, seaman with needed training courses served in these positions, i.e. the various gunners or men trained in the various weapons foreman courses. Note: S-35 was lost with all hands on February 28, 1942 northwest of Bizierta. All that was recovered during a search by S-54 and S-55 were some planking and its rubber boat; none of the crew were ever found.

Suited up in their life vests, these crewmen enjoy a break in the enlisted men's quarters located at the stern of the boat. They are wearing the leather sailors outfit and pea coats, at the right is a junior NCO as denoted by the fowled anchor sleeve insignia.

Comparative Rank Table

Kriegsmarine Rank	Translation
Seemann	Seamen
Matrose	Ordinary Seaman
Matrosen-Gefreiter	Able Seaman
Matrosen-Obergefreiter	Leading Seaman
Matrosen-Hauptgefreiter	Leading Seaman (4.5 years service)
Matrosen-Stabsgefreiter	Senior Leading Seaman
Matrosen-Stabsobergefreiter	Senior Leading Seaman
Unteroffiziere ohne Portepee	**Junior NCOs**
-maat	Petty Officer
Ober-maat	Chief Petty Officer
Unteroffiziere mit portepee	**Senior NCOs**
Bootsmann	Boatswain
Stabsbootmann	Senior Boatswain
Oberbootsmann	Chief Boatswain
Stabsoberbootsmann	Senior Chief Boatswain
Offiziere	**Commissioned Officers**
Fahnrich zur See	Midshipman
Oberfahnrich zur See	Sub-Lieutenant
Leutnant zur See	Lieutenant (Junior)
Oberleutnant zur See	Lieutenant (Senior)
Kapitaleutnant	Lieutenant-Commander
Korvettenkapitan	Commander
Fregattenkapitan	Captain (Junior)
Kapitan zur See	Captain
Kommodore	Commodore

Note: The title of Kommodore was not only used as a rank but as a position.

Enlisted and Junior Non-commissioned Officer Insignia

Career insignia were worn on the upper left sleeve of the uniform, which were produced in three ways: machine embroidered, hand embroidered and metal attached to a cloth backing for junior NCO's. The metal junior NCO career insignia tended to be worn only on formal dress uniforms. Junior NCO's career insignia were as enlisted men's, only superimposed over an anchor. The embroidered threads were of a yellow-gold color. Summer insignia were blue embroidered on a white background. Enlisted career insignia were also produced as a combined piece of insignia whereupon the career symbol was embroidered to the same piece of cloth as the chevron.

There was no tan backed tropical insignia for career insignia so the blue backed or summer white insignia were worn, sometimes being trimmed down before being sewn on for a better appearance. The following is a listing of career insignia worn by the Schnellboot crews:

1. Seaman - the five pointed (sea star), Junior NCO's - a plain fouled anchor.

2. Machinist - a six spoked gear cog.

3. Coxswain/Helmsman - two crossed anchors.

4. Signalman - two crossed flags, red with square white center.

5. Radioman - a pointed tipped lightning bolt.

6. Torpedo Mechanic - a torpedo superimposed upon a gear cog.

7. Mine mechanic - a mine superimposed over a gear cog.

8. Medic - snake wrapped around a pole.

9. Carpenter - a downward pointing compass.

   (Note: It is not certain how often medics or carpenters served on crews. However, this career designator has been seen on casualty and capture reports.)

This junior NCO (Machinist) wears his Schnellboot war badge and Iron Cross first class on his wedding day.

Enlisted and Junior Non-commissioned Officer Specialist Insignia

Specialist insignia were worn on the lower left sleeve of the uniform, below the career insignia. These were produced in two ways: machine embroidered and, less commonly, hand embroidered. The stitching was of a bright red color over dark blue or white cloth for summer uniforms. Chevrons denoted the number of training courses the sailor received. Some specialist insignia were redesignated in August 1940.

Specialist insignia worn by men of the Schnellboot crews are listed as follows:

1. Gunner/Gun Chief of automatic anti-aircraft weapons for small ships - flaming shell with one chevron.

2. Gunner/observer of automatic anti-aircraft weapons - winged shell.

3. Gun Chief of light anti-aircraft weapons - winged shell with one chevron. This insignia was introduced in August 1940.

4. Torpedo Control Foreman - an upward pointing torpedo. Grade III was with one chevron; grade II with two chevrons.

5. Electric Motor Specialist - slightly turned red gear cog with lightning bolts emanating from it. From 1933-1940, there were three courses. Course III had just the cog. Course II had the cog with one chevron, and course I had the cog with two chevrons. The Kriegsmarine introduced changes in August 1940. Course III with just the gear cog was changed to coastal searchlight course. Course III now had the gear cog with one chevron. Course II had the gear cog with two chevrons.

6. Motor Course - From 1933, a single three bladed propeller for motor course III. Propeller with one chevron. Motor course II had the propeller with two chevrons. The solo propeller insignia was being discontinued.

7. Blocking weapons Specialist/Foreman - a mine by itself until 1940, when two grades were introduced. Blocking weapons specialist, mine with two chevrons. Another change occurred in February 1942, when the blocking weapons specialist foreman were to receive the same badge as the blocking weapons specialist, namely a mine with one chevron.

Junior NCO (Machinist) in the pea coat.

Enlisted man wearing the dark blue insignia on the sleeve of his tropical shirt. The badge on his left pocket is for physical sports performance.

The four production styles for Junior Non-commissioned Officer's career insignia. Top left: Blue thread machine embroidered on white for the summer uniform; sometimes appearing on the machinist work jacket. Top right: Gilt metal version usually found on dress uniforms or on the peacoat. Lower Left: Machine embroidered. Lower right: Hand embroidered.

An assortment of machinist career and specialist insignia.

Examples of sailor's insignia of enlisted and junior NCO's; the sextant patch was worn by a navigator.

An assortment of enlisted and junior NCO radio operator and signalman sleeve insignia.

A Kriegsmarine signalman wearing his patch, devoid of any rank insignia.

This sailor has had his sleeve insignia trimmed down for wear on the tropical tunic.

These three portrait photos show the various weapons career insignia in wear. Top two: Torpedo mechanics. Bottom left: A mine or blocking weapons technician.

An assortment of torpedo related insignia. At the top, a hand embroidered version is tagged by the manufacturer as an example; in the bottom row, the silver chevron beneath the rank chevron denotes an NCO candidate. To the left are the white backed summer uniform versions.

An assortment of mine technician and gunner insignia. The top row being career insignia; the middle and lower row show the red colored training course insignia. The two chevrons are for scale.

This senior NCO wears the second pattern schnellboot war badge below the top left breast pocket.

Senior Non-commissioned Officer And Officer Shoulderboard Insignia

Senior non-commissioned officers displayed their rank on dark blue woolen shoulderboards with the appropriate colored trees and silver pips. Their career insignia were positioned centered on the shoulderboard and consisted of a gilt colored fouled anchor with the career emblems superimposed above the fouled anchor. Tropical issue shoulderboards for non-commissioned officers were tan colored cloth with a medium blue trees with appropriate career emblem and pips.

Officers shoulderboards were constructed of matt or bright colored cords sewn to dark blue or white for the summer uniform underlay with gilt metal pips further designating higher ranks. Career insignia in minature were placed centered on shoulderboard. Most Schnellboot officers were line officers with no career insignia on their boards. All shoulderboard insignia were attached to the uniform either by being sewn or slipped on through a cloth loop and secured by button or passed through a small opening at the seam of the uniform and secured by button.

This junior officer (lieutenant) wears his shoulder boards on a typical work shirt. The boat is of the S-30 class.

Top row: Officer of the line shoulder boards; the one on the far left is the sewn in type. Middle: Variations of the Sea Star i.e. officer of the line hand embroidered career cuff insignia; these were positioned above the sleeve rings. Bottom row: 3 types of Senior NCO Shoulderboards. Left: Torpedo mechanic. Middle: Engineering. Right: A tropical pattern line NCO slip on the shoulderboard.

The Schnellboot War Badges

Until May 1941, German Schnellboot crews received the Destroyer badge for combat recognition. Documents showing this have the flotilla number.

The requirements to receive the Schnellboot war badge were as follows:

I. General requirements: worthiness and good conduct.

II. Special requirements:

(a) a minimum of twelve missions against the enemy.

(b) one highly successful mission, or the individual soldier showing exemplary action, or is killed, or loss of the ship to the enemy and, in some cases, being wounded in action.

III. Other ways of earning the badge: having been killed in action, or death caused by wounds or accident while on a mission, but completing or nearly completing the requirements listed under IIa.

IV.The leader of Torpedoboote/Schnellboote has the power to expand the requirements.

There are two types of Schnellboot war badges. The first was drawn by designer W.E. Peekhaus of Berlin, and was approved by the Kriegsmarine on June 12, 1941. It had an oak leaf wreath, with the national eagle on the top and a type 30 Schnellboot plowing through the waves at speed. Note the portholes on the side of this boat as well as the upswept hull sides above the torpedo launching tube. Considering only sixteen boats of this type were made, a second badge was initiated by Fuhrer der Schnellboote Petersen and designed by Peekhaus. This was slightly larger, featuring an S-26/38 type boat. This change represented the majority type of schnellboots produced at the time. In late 1943, the armored bridge boats began to appear. However, no further type of badge was produced.

A rarely awarded Schnellboot war badge with diamonds was presented to eight officers, Knights Cross holders of the Schnellboot forces. The badge was only made in the second pattern type and had diamonds filling in the swastika below the national eagle.

The Schnellboot war badge in wear by this junior non-commisioned officer.

A grouping of various tropical (Mediterranean) insignia and awards. On the left is a Naval award certificate, for the Afrika campaign cuffband, given to a member of the 7. Schnellbootsflottille. To the right, is his Schnellboot war badge certificate. Below are two versions of the Afrika cuffband; on the top is the Naval pattern and below is the standard or Army pattern in tan.

Schnellboot Insignia
the Early Years
1932-1939

Before the outbreak of World War II, the Schnellboote showed their number designators on both sides of the bow. There are a number of combinations seen in the photographic record. These combinations are as follows:

1.   Large white number on both sides of the forward bow.

2.   Large black number with a smaller s and number above it on both sides of the forward bow.

3.   Large white number with a smaller s and number in black above it on both sides of the forward bow.

The early boats carried large cast national eagle emblems on both sides of the bridge. Later in the war, remaining boats relegated to service in Norway are seen with the insignia of the German coastal security forces (Tigerverband), which featured a roaring tiger's head superimposed over two crossed swords. Boats, relegated to training purposes later in the war, had capital case letters or numbers painted on their bow. These did not necessarily pertain to the boat's number. Ironically, after the war, bow numbers reappeared on the bows of the surrendered Schnellboote. This time it was for the purpose of allied control.

Two examples of boats with black painted bow numbers.

Left: S-12 has additional applied smaller numbers above its bow rub rails.

Below: The Tigerverband insignia of the German coastal security forces in occupied Norway.

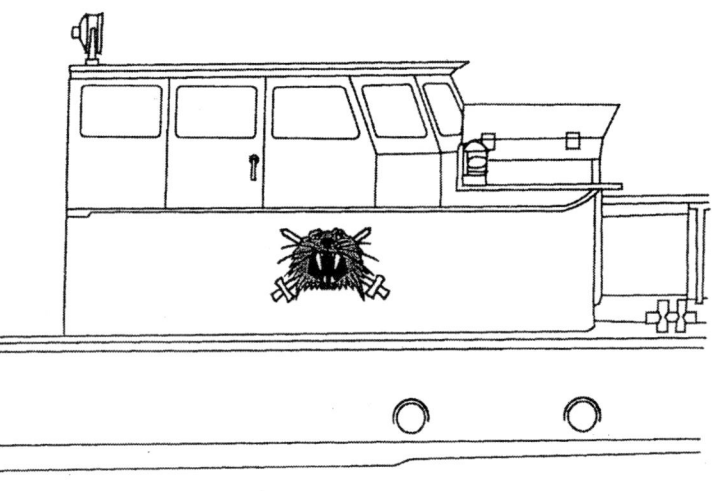

The 1. Schnellbootsflottille

Formed in mid 1935, when the 1. Schnellbootshalbflottille received enough boats to constitute a full flotilla. The Flotilla saw action near Gotenhafen during the invasion of Poland then in the North Sea. They participated in the invasion of Norway, conducted operations in the Dutch and French waters as well as the English Channel. In May 1941, the flotilla conducted operations in the Baltic Sea. Early '42 the flotilla was transferred to the Black Sea and conducted operations against the Soviet Navy. When the Eastern front collapsed, the flotilla ended its operations there in October 1944, but was reformed in late '44. The flotilla saw action in the Baltic to the end of the war. Boats of the 1. Schnellbootsflottille are seen with a tiger (S-24), and in the Black Sea, at least one boat had a pouncing panther. However, these photographs are not very clear, and the boat's number is unknown.

Top: S-24 with its tiger insignia and tonnage sunk number beneath. Also of note is the large cast Eagle national symbol behind the tiger, a feature found on most early war boats. (Photo courtesy of D. Krakow) Below: A 1. Schnellbootsflottille crew photo. Front row, fourth from right is German cross in gold recipient Machinist Mate Otto Enders who carried out 150 missions. Also of interest are the Krim Shields worn on the upper left sleeves of the two sailors on the far right of the image. This award denotes their service with the flotilla in the Black Sea.

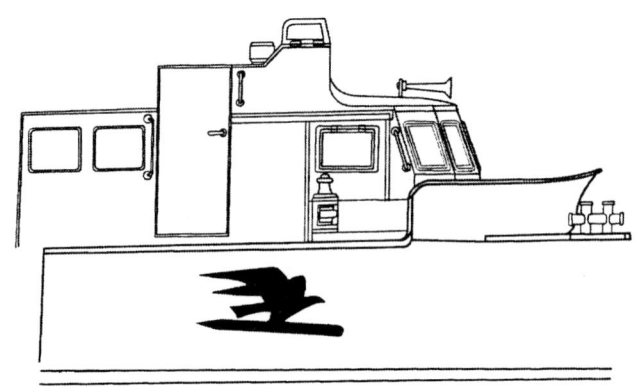

Left: a Popeye like cartoon adorned the open bridge splash guard on this 2. Flotilla boat. Right: a version of the eagle carrying a torpedo, this version being highlighted and colorized; the illustration below shows the all black version. (Photos courtesy of H. Burger)

The 2. Schnellbootsflottille

Formed in August 1938, the 2. Schnellbootsflottille saw action in the North Sea, conducting operations aiding the invasion of Norway, after which they operated in the southern part of the North Sea and English Channel. In May 1941, the flotilla operated in Finnish waters. After November 1941 and to the end of the war, the 2. Schnellbootsflottille saw action in the English Channel and southern part of the North Sea. Early on in the war this flotilla's boats make little use of insignia. Some time after they had left Finland, a system of playing card symbols was introduced. The Ace symbol with a rectangular dash under the symbol was the lead boat of each pair (Rotte). Allegedly, (not confirmed), boats 9 and 10 made use of the joker symbol. At least one boat had the entire outline of a playing card painted on its bulwarks. Also observed are smaller Ace cards painted on the bridge splashguards above the wheel house.

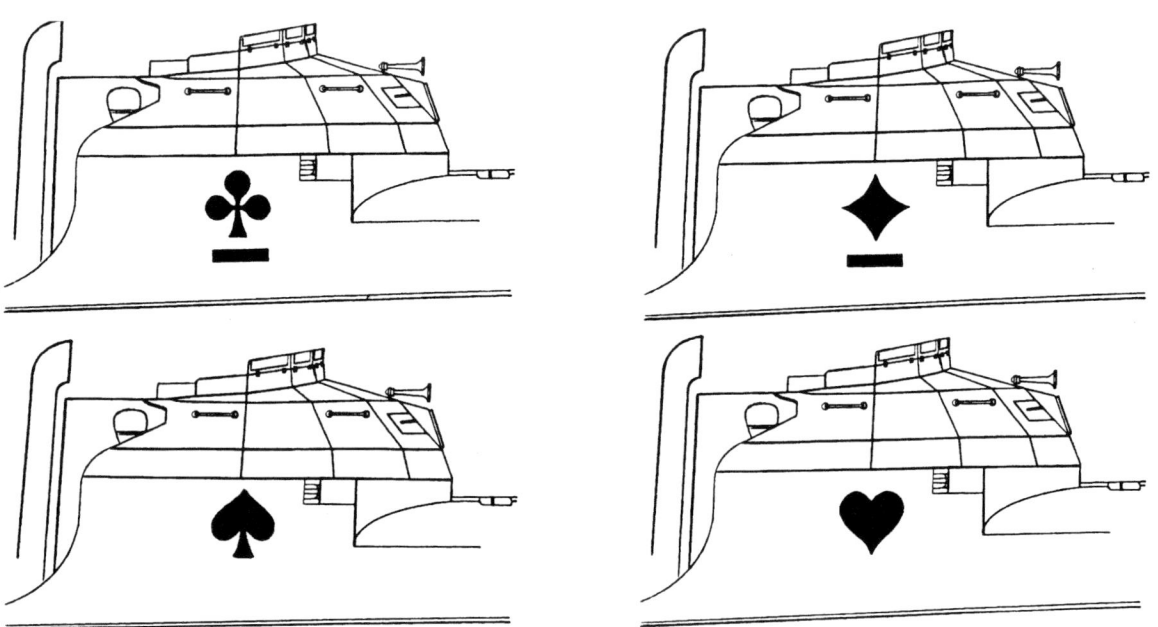

While operating out of Finland, crews of the 2. Schnellbootsflottille wore small pins of the Finnish Flag on their side caps or black handkerchiefs. (Photos courtesy of H. Burger)

A candid shot of an unidentified 3. Schnellbootsflottille boat commander at the open bridge of a type 30 Schnellboot.

The 3. Schnellbootsflottille

Formed in Kiel in May 1940, the 3. Flotilla consisted of only the type 30 Schnellboote. This flotilla saw action off of the Dutch coast and English Channel from 1940 to May 1941. In the fall of 1941 the unit conducted operations in the eastern Baltic Sea (operating from Pillau, Windan, Liban, Riga and Turku). In November 1941 the flotilla was moved down the Rhein and Rhone Canals, disguised as work boats with the crews wearing civilian (i.e. barge operator) clothing for security reasons. The flotilla operated in the Mediterranean Sea off the North African Coast, Tunisia, Bizerte, Sicily, Napels, Torlon and Marseille. In November 1943 the flotilla moved from the Tyrrhenian Sea to the Adriatic Sea until the end of the war, surrendering with the remnants of the former 7. Schnellbootsflottille in Ancona. This flotilla made use of black painted aquatic sea life. Due to repainting, sometimes the boats are to be seen void of any insignia; also with the repainting, slight variations exist. Not all boats were identifiable. It is not known, for example, if the arched back crocodile of S-54 is just a variation of the lying down version of the crocodile. This flotilla made wide spread use of tropical issue shirts with Luftwaffe pattern eagles over the right breast pocket. In June 1943, near Tobruk, a large British 8th Army supply depot was captured by German troops, and the men of the flotilla obtained wool overcoats which were put to use on cold nights.

Top: Photos of S-31 with its flying fish insignia are rare, having been sunk off Malta on May 10th, 1942 by a mine. Below: An unidentified 3. Flotilla boat with a sea horse insignia.

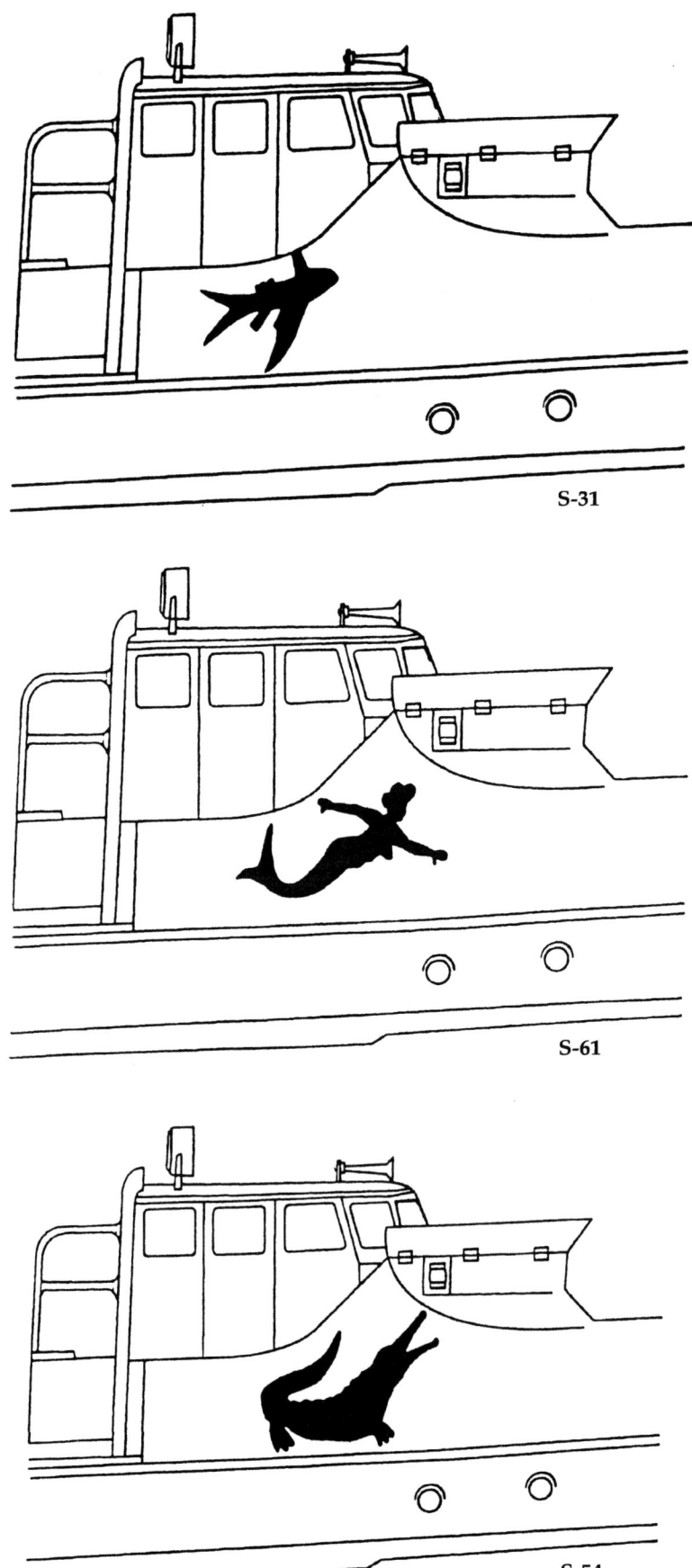

S-31

S-61

S-54

Unknown

Unknown

Unknown

Unknown

Unknown

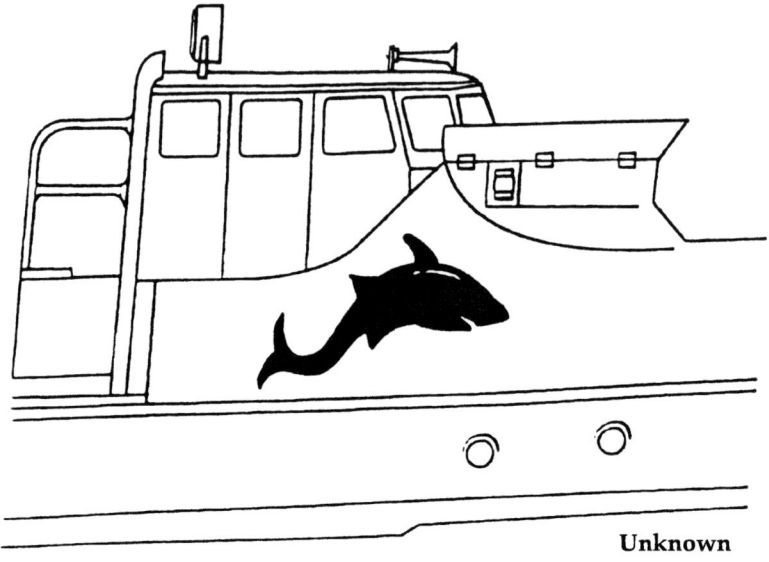

Unknown

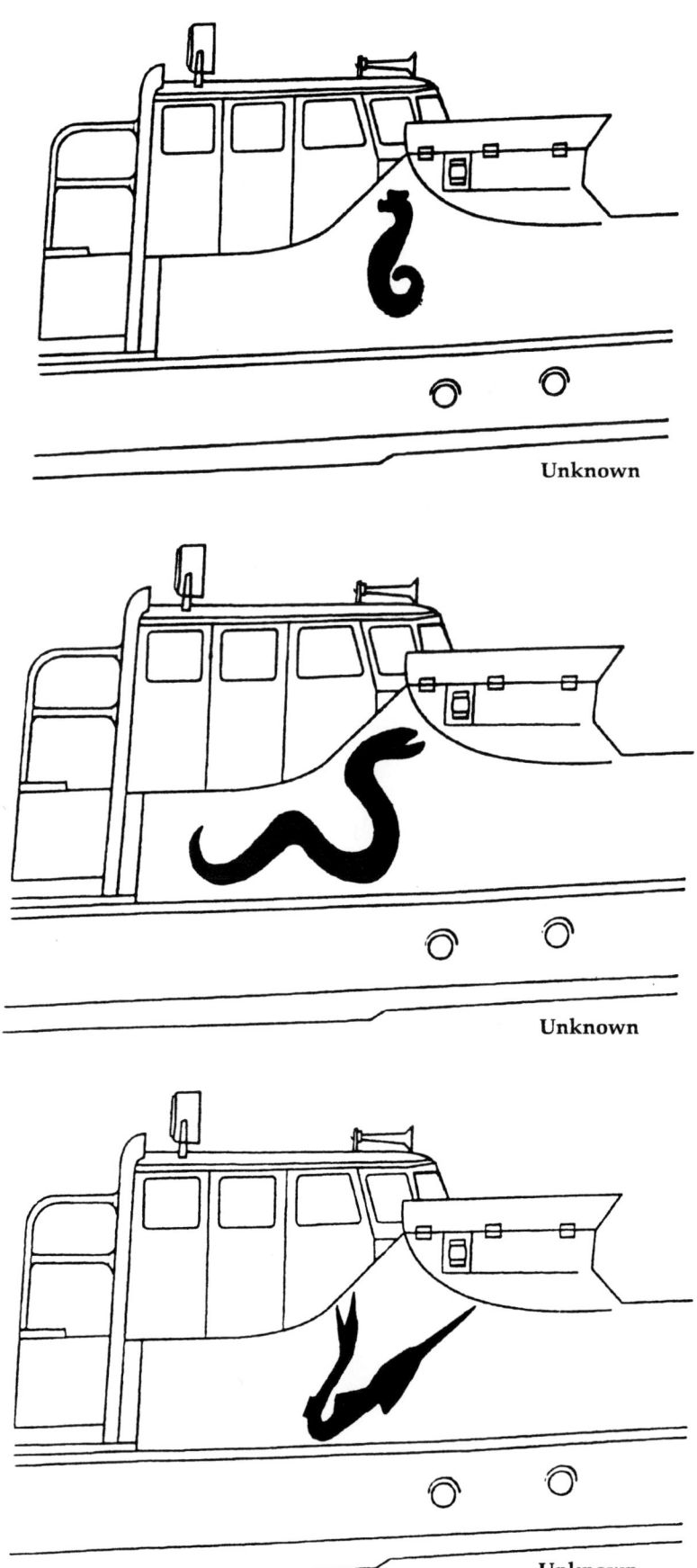

Unknown

Unknown

Unknown

Two famous Schnellboote from the 3. Schnellbootsflottille. Above: S-54 with its open mouth crocodile. Below: S-61 with its mermaid insignia. It was these two boats which captured the city Venice without firing a shot after Italy joined the alliance in 1943.

Funkobergefreiter (S-31) Heinz Hildebrand wears the metal cockade and eagle of the sailor's cap on his white board cap. The photographic record shows this practice was widespread within the 3. Flotilla while in the Mediterranean. Below: An unmarked 3. Schnellbootsflottille boat off the north coast of Africa.

The 4. Schnellbootsflottille

Formed in October 1940, the flotilla saw uninterrupted action in the North Sea and English Channel. After D-Day 1944, they operated out of Rotterdam, Holland. The flotilla's insignia, which was an open jawed black panther, was located on both sides of the hull at the bridge. Some of the type 100 boats also made use of their commander's name being painted on the armored bridge. An assortment of stern insignia is also in the photographic record on S-100 type boats of this flotilla.

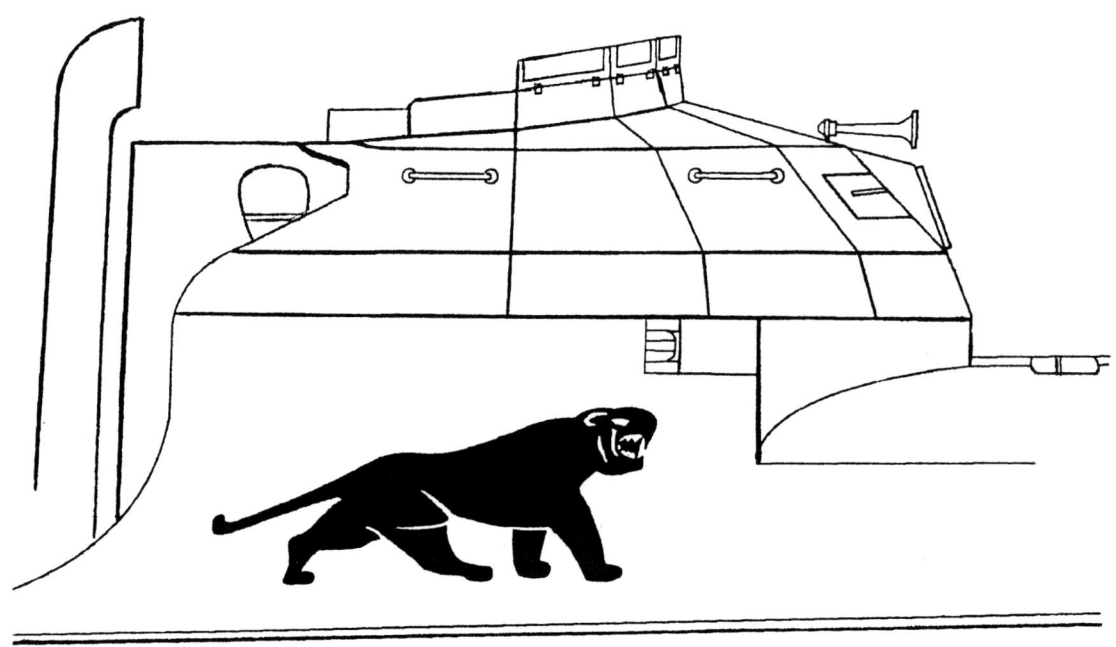

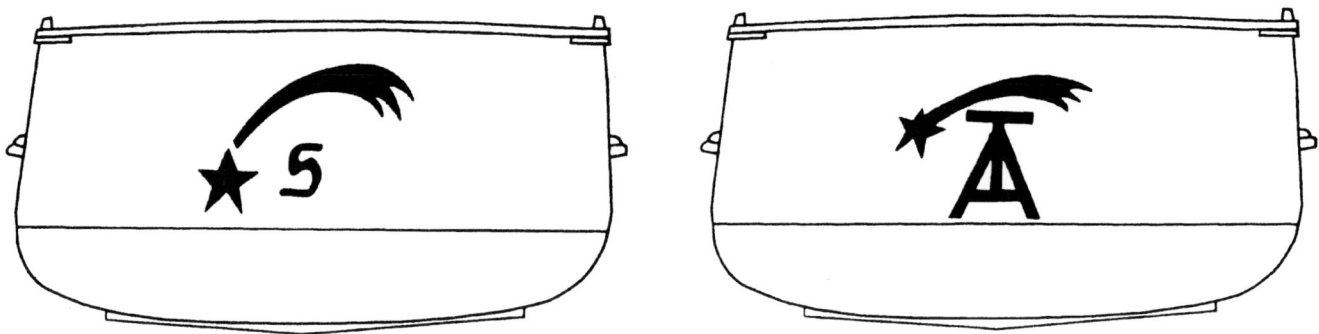

Above: Two examples of stein insignia of the 4. Schnellbootsflottille.

THE 5. SCHNELLBOOTSFLOTTILLE

Formed in July 1941, this unit saw action in the Baltic and English Channel until being destroyed in the course of the Allied invasion (D-Day). The flotilla reformed in July 1944, operating in the Baltic out of Helsinki. After evacuating Finland in September 1944, the flotilla operated out of Reval, Windau and later Gotenhafen. At the coming of the new year, it operated in the North Sea and English Channel. In April 1945 the flotilla was transferred to Swinemunde, then to Bornholm which is based in Denmark and later Laboe. It's action in the Baltic, involved the evacuation of the troops trapped in the Kurland pocket.

THE 6. SCHNELLBOOTSFLOTTILLE

Formed in March 1941, this unit saw action in the North Sea and English Channel. For a short time in 1942 it operated out of Norway. Operations in early summer 1944 were in Finland. After the D-day invasion and up to the end of the war, the Flotilla conducted operations in the English Channel. Boats of this Flotilla had shields with a sword superimposed, flanked by two stars on each side of the sword. A Latin inscription was on the top edge, but it is not legible in known images of the period.

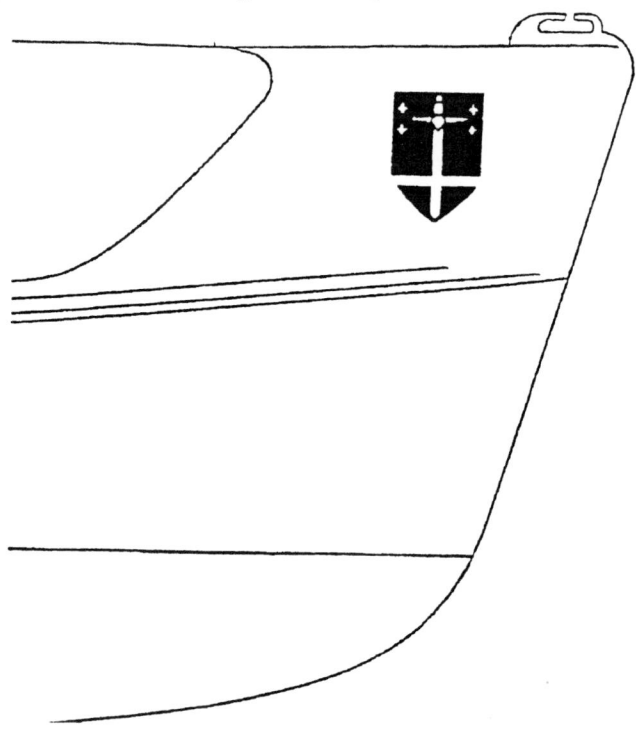

These seemingly amused crewmen standing in the open bridge have small letter W's adorning their overseas caps. The pedestal like device with the binoculars mounted on top is the torpedo targeting system.

The 7. Schnellbootsflottille

Formed between October 1941 and April 1942, it operated until fall of 1942 in the Baltic Sea, and then transferred to the Mediterranean via the Rhein-Rhone canal system. There were eight boats total, S-151 to S-158. This type of boat was smaller than the 3rd Flotilla's S-30 type and could not carry out mine laying operations. In the Mediterranean, the flotilla saw action in the following areas: Tunisia, Sicily, the North African coast, Augusta, Palermo, Messina, Vibo, Valencia, Salerno, Nettuno, Toulon and Genoa. In early 1944, the flotilla operated in the Adriatic Sea. In October of that year, due to attrition, the 7th Flotilla combined with the 3rd Schnellboot Flotilla. At wars end, four boats from the original 7th Flotilla were left: S-151, S-152, S-155 and S-156. They were surrendered to English forces on May 3rd, 1945 at the Italian port of Ancona.

The 7. Schnellbootsflottille made use of capital case Latin letters for their boat's recognition. One boat is seen using a crossbow as an insignia in lieu of a capital case letter. The letters however, apparently did not correspond to their commander's last name as in other Flotillas. Some of the crews crafted small Latin letter cap badges corresponding to the boats they served on.

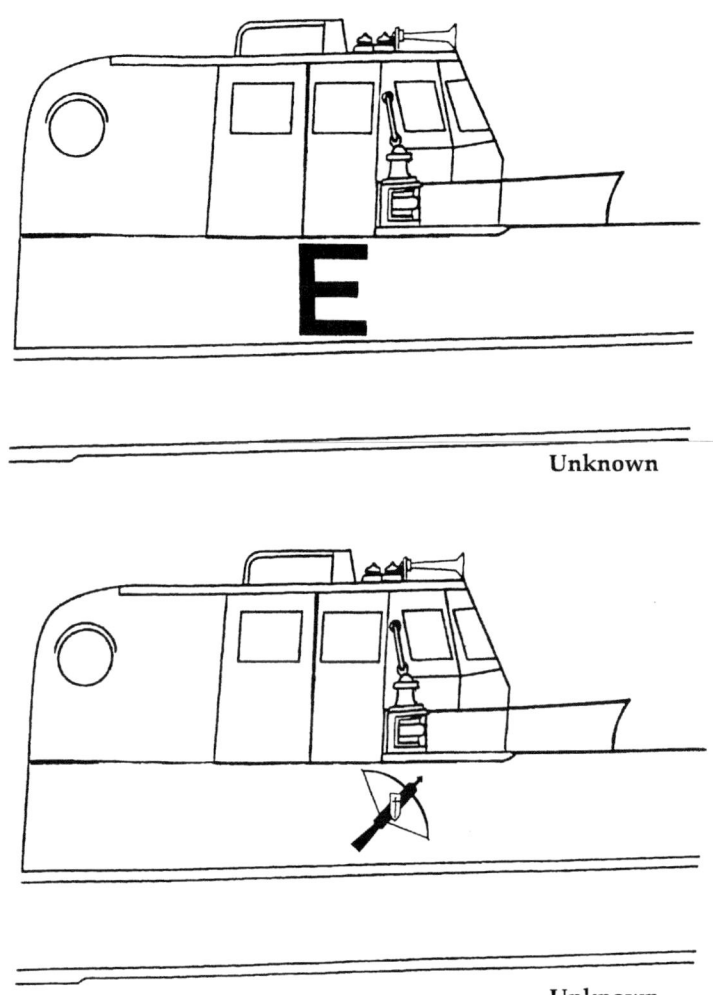

Unknown

Unknown

One unidentified boat from the 7. Schnellbootsflottille had a cross bow with a shield for its insignia. This was a deviation from the capital case Latin letters painted to the other seven boats of this flotilla.

Above: Three views of S-155 (W) in the Mediterranean. Note the small caliber hits on the hull in the top left photo.
Below: S-157 (C) and S-151 (H).

Unknown

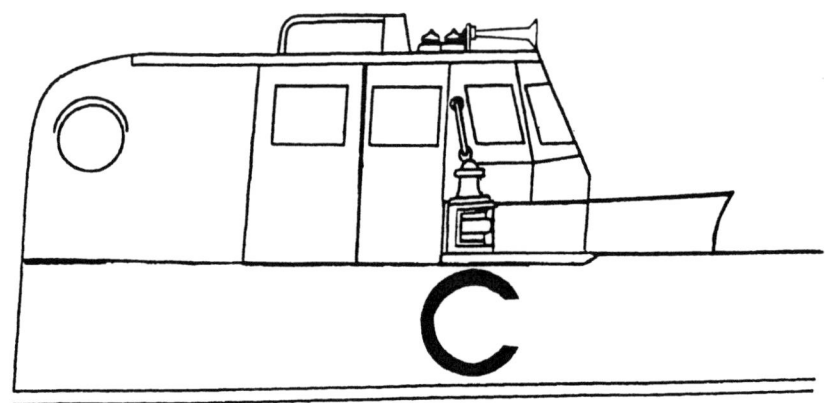

S-157

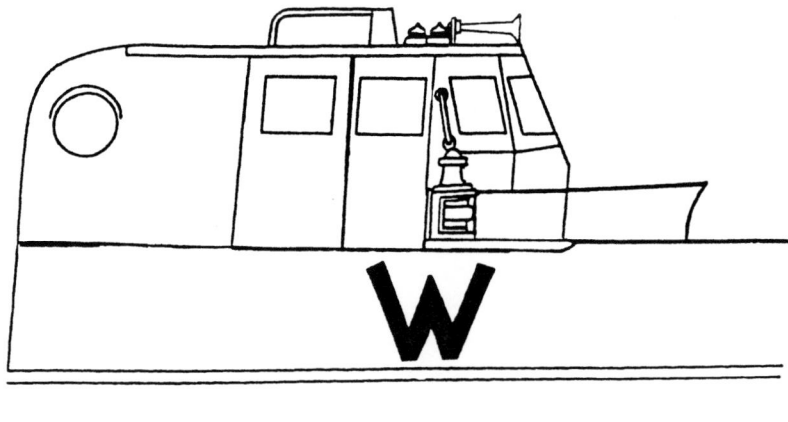

S-155

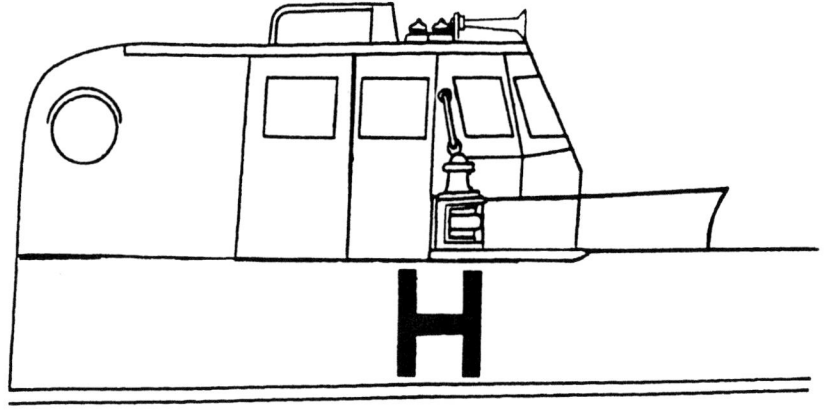

S-151

Unknown

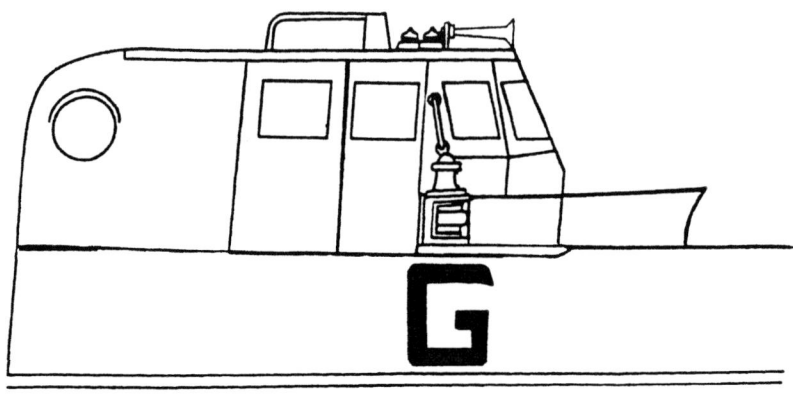

Unknown

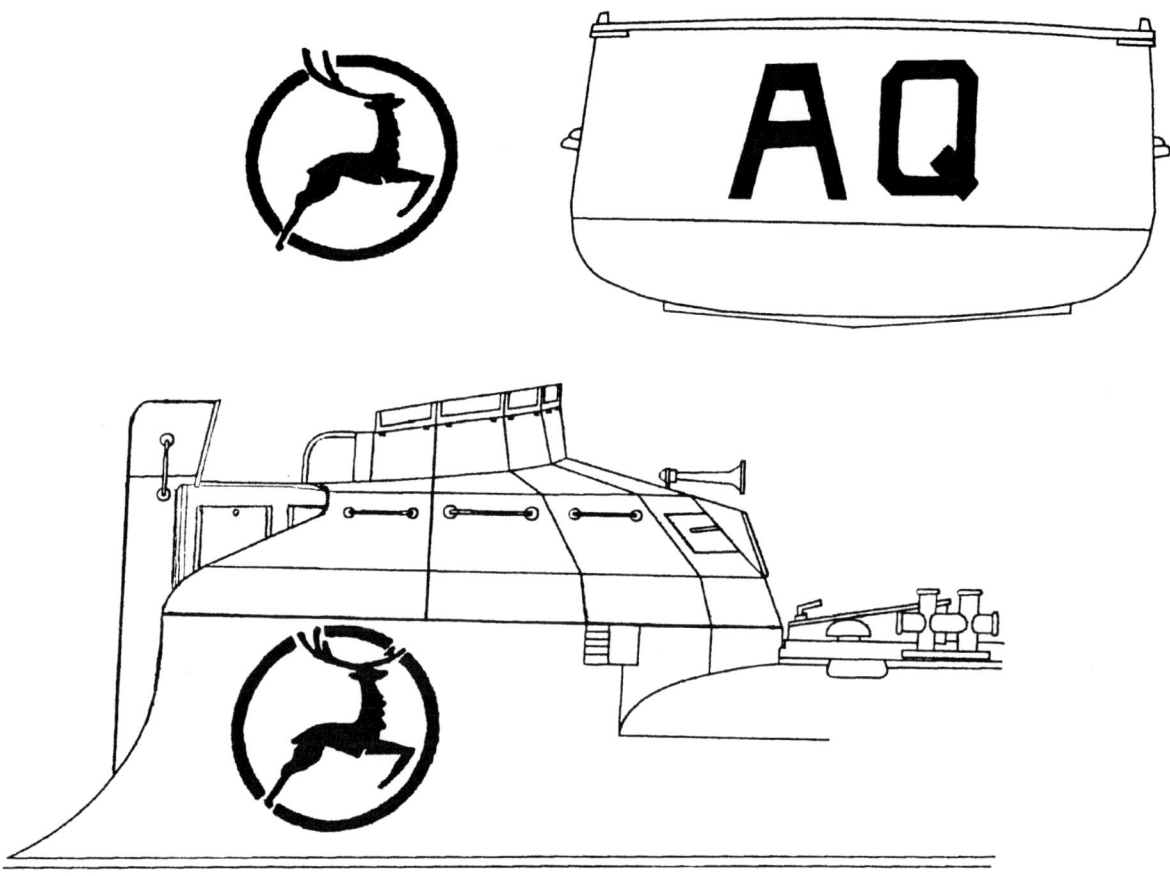

THE 8. SCHNELLBOOTSFLOTTILLE

Formed in November 1941, this flotilla conducted operations from various Norwegian ports. Later operations were carried out of the Port of Kirkenes with the intent of disrupting Soviet forces. The boats were found to be unsuitable for the extreme arctic conditions, and therefore the flotilla was disbanded in July 1942. It reformed in December 1942 and saw action off Norway, in the English Channel, the southern part of the North Sea and in early 1945, the Baltic. The Flotilla's insignia was a reindeer leaping through a circle. The reindeer was to commemorate their service above the artic circle (Norway). Two variations exist: one is with the horns protruding out of the circle without frontal horns; and the other style had front and back horns just breaking through the circle. The boats of this flotilla also painted the initials or first letter of their commander's name in stylized Latin lettering on the armored bridge.

A 9. Schnellbootsflottille boat possibly S-150 which had its bow ripped off on the night of June 13/14, 1944 during the massive air raid on the Port of Le Havre. Just visible is the shield and cross insignia, as well as an indiscernible boat's insignia on the armored bridge

The 9. Schnellbootsflottille

Formed in April 1943, this flotilla saw action exclusively in the English Channel. Boats from the 5. And 9. Schnellbootsflottille were involved in the Lyme Bay disaster. On the night of April 27/28, 1944, two LSTs were torpedoed and sunk, a third was badly damaged, with great loss of life. This incident was the catalyst for the increased bombing and development of bunker busting bombs that would hound the Schnellboot forces in the English Channel to the end of the war. The insignia used on the boats of the 9. Flotilla was a long shield with a cross; a smaller black Iron Cross was centered atop of the long shield.

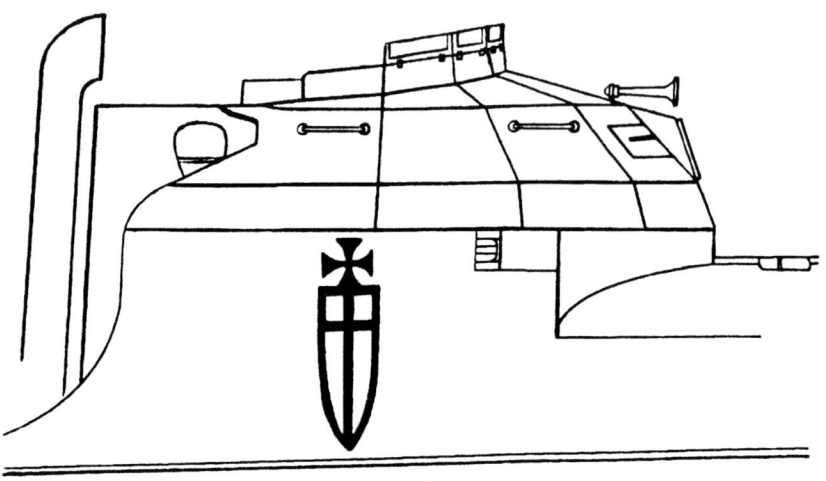

The 10. Schnellbootsflottille

Formed in March 1944, it saw action in the English Channel.

The 11. Schnellbootsflottille

Formed originally in February 1943, this flotilla never came to action, being formed and disbanded three times with available boats and crews being transferred as replacements for other depleted Schnellboot units.

The 21. Schnellbootsflottille

Formed in September 1943 in Eckernforde, it was composed of the small LS type Schnellboote. The flotilla was transported by train to the Aegean Sea and put under the command of the 1. Schnellbootsdivision. This flotilla was disbanded in October 1944.

The 22. Schnellbootsflottille

Formed in December 1943, after training, the flotilla was transported by train to Italy in May 1944. This unit's boats were of the small KS type (coastal Schnellboot). Croatian crews received training on the boats in Lignano. In October 1944, the boats were turned over to the Croatian Navy and the unit was disbanded.

The 24. Schnellbootsflottille

Formed in December 1943 and comprised of former Italian boats, the flotilla conducted escort and security duties in the Aegean sea until the evacuation of Greece. After a short time in the Adriatic Sea, the unit was disbanded in October 1944.

Schnellboot Tenders

With the exception of the Tender Nordsee (North Sea), Schnellboot tenders were named after either colonial explorers or places. The tenders had no specific individual insignia, with the exception of the Tsingtau which featured a shield shaped crest with an oncoming view of an early type Schnellboot. Smaller versions of the crests were affixed at the bows of the Tsingtau's life boats and launches. All of the tenders had large cast National eagles attached to their bridge structures as destroyers and some minesweepers did.

Tsingtau: in service September 26, 1934. It was named after the capital city of Germany's colony in China.

Tanga: in service January 19, 1939. It was named after a major city on the East African coast where the British attempted a landing in 1914. The battle was a German victory.

Carl Peters: in service January 6, 1940. It was named for the colonial explorer, Reichskommisar and Governor of East Africa.

Adolf Luderlitz: in service June 11, 1940. It was named for the colonial explorer and Governor of South West Africa.

Hermann von Wissman: in service December 26, 1940. It was named after a colonial explorer; Herman von Wissman who was a friend of Carl Peters. Wissman succeeded Peters as Governor of South West Africa.

Gustav Nachtigal: in Kriegsmarine service May 13, 1944. Nachtigal was a colonial explorer and Governor of South West Africa.

The bow crest of the Tsingtau. It's exact colors are unknown.

Note: The Buea (May 1944 to the end of the war), the Romania (1942 to 1944 in the Black Sea) and the Estonia (1941 in the Baltic), these were captured ships pressed into Kriegsmarine service, probably as a stop-gap measure. Little more is known about them.

The Tender Nordsee. Of note are the large cast eagles located at the sides of the bridge.

Two views of the Tsington. Note the large crest which adorned both sides of the ship's bow.

The Tanga

Flags and Pennants

Flags and pennants as used by the Schnellboot forces were of two types, official and unofficial. This section explores only the flags as seen on boats in the photographic record, not the numerous parade and higher command flags.

1. **1933, Reich war flag:** Had black, white and red colored horizontal bands superimposed by a black iron cross, slightly off center towards the left.

2. **1935, Reich war flag:** Brought into service on November 7, 1935, this flag was completely different from it's predecessor. Colors were red white and black, featuring a swastika near its center. The iron cross was in a red field in the upper left corner near the seam. Black and white bars were extended out from the circled swastika which was slightly offset towards the seam.

3. **1935, Reich war flag, Second pattern:** This flag was the same as the first 1935 version, except the swastika was slightly more offset to the left and the black rings around the swastika were connected to the smaller black arms which were emanating out of the circled swastika.

4. **Flotilla Command Pennant:** A white triangular pennant with a black, and black bordered, iron cross.

5. **Captain's Pennant:** A very long white pennant with a deep V cut into it like a swallow's tail. This flag had a small black with black bordered iron cross near the seam and was flown above the war flag.

6. **Hand signal flag(s) / Semaphore:** A red flag with a white square or rectangular center, this was the letter "C" (Casar) in the German Semaphore alphabet. Two of these were wielded by the signalman, with a particular pattern representing a letter. A special platform was located aft the bridge to get the signalman as elevated as possible.

7. **Award pennant for downing an enemy plane:** Brought into use by the Kriegsmarine in the mid 1940's, it was to be flown below the war flag. This pennant was 35 cm long with a black framed, black iron cross inside of a white circle, with the remainder of the pennant being red. The end tip of the pennant was rounded.

8. **Victory Pennants:** These pennants were unofficially used to celebrate the sinking of an enemy vessel or vessels. Often hand made on board, they were flown for limited time periods often affixed to one of the various radio antennae. The practice was prevalent in the U-boat forces. The victory pennants were small, of varying sizes, usually white with the tonnage of the ship that had been sunk hand-written in the center of the pennant. The lead boat of the 3. Schnellbootflottille flew a red pennant above the English naval flag to denote the victory of the capture and sinking of ML-130 near Malta in May 1942. Both flag and pennant were affixed to the short wave antenna.

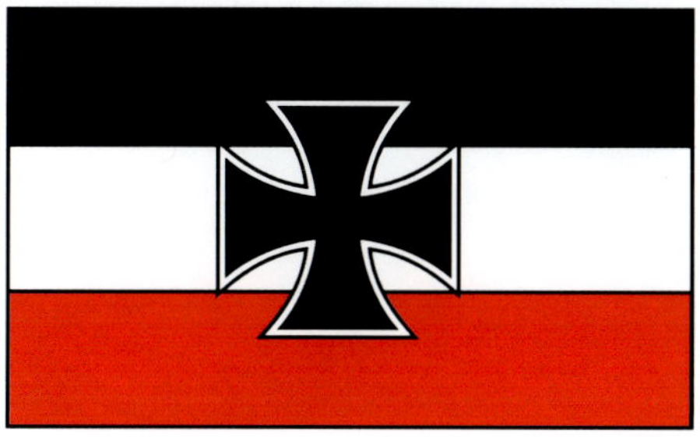

(1.)

(2.)

(3.)

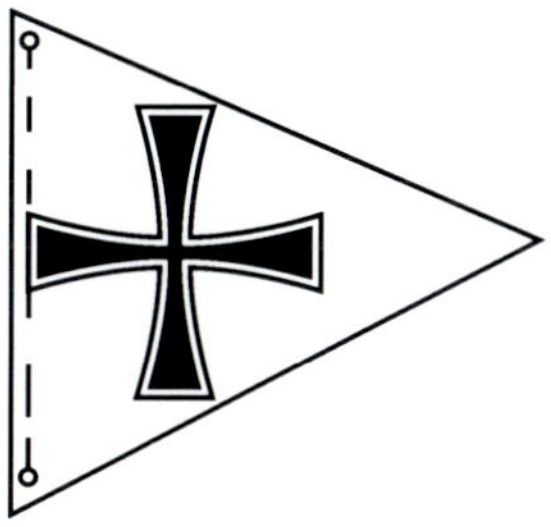

(4.)

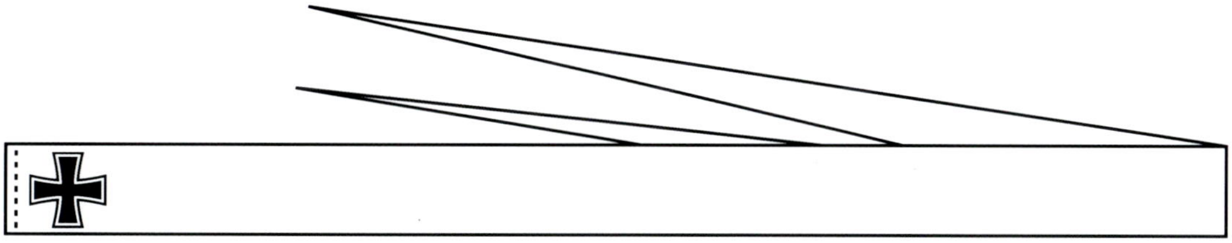

(5.)

(6.)

(7.)

(8.)

NOTE: Flags and Pennants are not drawn to scale.

1935 Reich war flag, being flown with the streamer-like captain's pennant.

This enlisted sailor is perched on a railing of the Tsingtau, astride one of the boats life rings.

Life Rings

Prewar and early war life rings were red/orange, often with the Flotilla or ship's name in Gothic style white lettering. Later due to security, this practice was abandoned, giving to all red/orange or intermittent (by quarters) red/orange, white (see photo on back cover). Some period photos have the ship's number and/or flotilla actually censored (blotted out) of the photograph as seen in the photograph on the right.

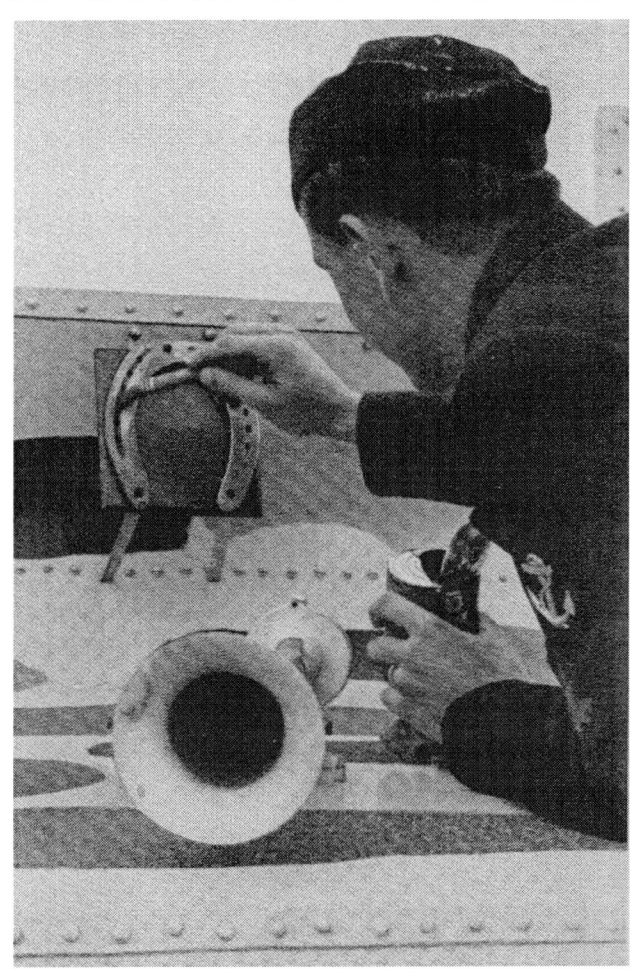

Talisman Traditions

The Germans usually bore gold painted horseshoe talismans, with the opening facing downward, unlike the British, who believed this position would allow the "luck" to fall out. The horseshoe talisman is seen in period photos, also affixed to German trucks and even tanks. (Above photos courtesy of Hugo Burger)

Machinists of all ranks took part in specific diesel engine training by the firm Daimler-Benz. It can be speculated the small pin, propeller badges seen in some photos were obtained by crewmen and worn as a sign of pride.

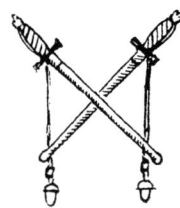

Officers of the Naval Academy graduating class of 1937 sometimes painted two crossed officer's daggers on the upper splashguard or on the side of the armored bridge house.

PHOTO ALBUM

This final chapter provides a number of candid shots of the Schnellboot personnel at work, at sea or dockside. Various uniforms and work apparel are seen to be worn. There are also photographs of unidentified boats, a number of which have various animals painted clutching torpedoes. No drawings have been attempted, however, due to the unclear imagery.

Missions in the English Channel required heavy leather over garments. The crew above listen to a briefing by their commander.

Training boats with their letters on the forward bows. They are moored alongside one of the tenderships.

May 1945, a sad end for the remnants of the 3. and 7. Schnellbootflottille, here surrendering to the British at the harbor town of Ancona, Italy. These men would face years of internment, some as far away as Egypt. (Photo courtesy of D. Krakow)

Top Photos: The black leather coats, with and without collar. Below: Overcoats being worn on deck. Note the use of the belt and buckle.

Above: Two crewmen at the furthest edge of the stern wearing wool knit caps. Below: A heavily frosted-over wheel house on a type 38 boat.

Above: The wheel house and open bridge of a type 38 boat. Below: The armored bridge of a type 100 boat which appears to be tied alongside one of the tender ships.

Top: A pristine condition 1935 Reich war flag, flown midships on S-31.

Below left: Close up view of the stern smoke dischargers on a type 30 boat.

Below right: S-26, after a bad engine room fire, made it's way to Linz for repairs, returning back to the 1. Schnellbootsflottille on 8 September 1942.

Above: A type 38 in a floating drydock. Below: A view of the Tsingtau. Of note is the 1933 Reich war flag being flown from the stern.

Only 8 of the S-151 type Schnellboote were produced. This page and the following three pages show the crews and their boats in action. Below: Two crewmen pose astride the port torpedo tube.

Top: A starboard mounted MG-34. Below: Oblt.z.See Heckel, commander of the S-155, sitting on one of S-155's ventilator trunks with his wounded lead engineer. Right: Aft mounted 20 mm cannon.

Above: The open bridge of S-155 looking aft. Commander Oblt.z.See Heckel is to the left. Below: Stern view. Note the white painted helmets, racked on the aft gun station platform. Present day militaria dealers would consider these to be fake, because they are not the usual gray green color.

Top: Upkeep, painting and maintenance were constant in the Schnellboot forces. Below: View from the stern looking forward on an S-151 type. Of note is the upgrade to a twin 20 mm cannon.

Top: Two views of an unidentified boat with what appears to be a mosquito carrying a torpedo. Below: Another boat with what may be a fox carrying a torpedo. (Photo courtesy of K. Scheuch)

The crew of this unidentified 38b type Schnellboot appear to have made use of the imperial eagle design as found on the 1866 medal for the battle of Koniggratz.

ABOUT THE AUTHOR

German-American, Richard W. Mundt has been an armor and schnellboot enthusiast since his teens. He lived in Germany for ten years, four of which were spent serving the United States as a tank crewman on M60A1 and M60A3 tanks. He is a scale model ship builder and militaria collector. Mr. Mundt reads and speaks fluent German and has had many friends who served Germany during World War II.

WEBSITES

www.germanschnellboot.com

www.crimsonbooksinc.com

BIBLIOGRAPHY

Auszeichnungen des Deutschen Reiches 1936-1945, K.G. Klietmann (Motorbuch Verlag 6. Auflage 1991).

Die Afrika Flottille, Konteradmiral A.D. Friedrich Kemnade (Motorbuch Verlag 1. Auflage 1978).

Die Deutschen Schnellboot im Zweiten Weltkrieg, Gerhard Hummelchen (Mittler Verlag 1996).

Die Kriegsmarine, Uniforms and Traditions, Vol. I, II and III, John R. Angolia and Adolf Schlict (Bender Publishing 1993).

Die Kriegsmarine (magazine) May 1944, Berlin

Schnellboot In Action, Warships Number 18, David L. Krakow (Squadron Signal Publications).

Schnellboote in Einsatz 1939-1945, Volkmar Kuhn (Motorbuch Verlag 2. Auflage 1986).

Schnellboote Vor!, Hugo Burger (Gerhard Stalling, Verlag 1943).

For other books in this series and to see
our full line of military and historical books please visit our website:

www.crimsonbooksinc.com